THE BAT DETECTIVE

A Field Guide for Bat Detection

Brian Briggs

and

David King

Stag Electronics
Steyning, West Sussex

© Brian Briggs and David King

Cover design based on drawing by Rowena Varley

The rights of Brian Briggs and David King to be identified as the authors of this work have been asserted by them in accordance with the Copyright, Designs and Patent Act 1988.

First published in Great Britain 1998
by Stag Electronics
15 Sir Georges Place, Steyning, West Sussex BN44 3LS.

All rights reserved. No part of this publication may be reproduced, stored in a retrieval system, or transmitted, in any form, or by any means, electronic, mechanical, photocopying, recording, or otherwise, without the prior permission of the authors.

ISBN 0 9532426 0 9

CONTENTS

INTRODUCTION	1
ACKNOWLEDGEMENTS	3
ABOUT BATS	4
RIGHT FROM THE BEGINNING	6
WHY DO THEY FLY LIKE THAT?	9
WING SHAPE AND BEHAVIOUR	11
PHYSICS OF SOUND AND THE JARGON	14
DESCRIBING BAT ULTRASOUND	18
VARIATIONS IN A CALL	21
METHODS OF DETECTING ULTRASONIC CALLS	24
INTERPRETING THE SOUNDS OF THE HETERODYNE DETECTOR	26
USE HEADPHONES	29
THE ZERO POINT	30
INDIVIDUAL SPECIES ACCOUNTS	34
FORAGING STYLES	46
GETTING STARTED	47
A BAT DETECTOR SURVEY	49
SUMMARY CHARTS	51
BAT DETECTOR SUPPLIERS	52
USEFUL ADDRESSES	52
INDEX OF CD TRACKS AND COMMENTARY	53
FURTHER READING	56

INTRODUCTION

Over the past few years, there has been an enormous growth in the interest shown in the use of bat detectors for survey work. The First European Bat Detector Workshop held at Gorssel, in the Netherlands, in 1991, demonstrated the techniques that were being used to produce a distribution atlas of bat activity for the whole of the Netherlands. The methods learnt there were shared at the first UK Bat Detector Workshop, organised by The Bat Conservation Trust, on the Isle of Wight in 1992. With modifications, those skills and techniques have since been taught, at many similar workshops throughout the UK.

Identification of bats in flight and the interpretation of their behaviour requires many skills and the use of as many different clues as possible: habitat usage, flight style, hunting technique, ultrasonic call, morphology etc. This book attempts to provide an introduction to these techniques in an accessible style, with the minimum of technical jargon.

The availability of affordable, reliable and accurate heterodyne bat detectors has made effective, nocturnal field work possible and the book naturally places considerable emphasis on the understanding and interpretation of bats' ultrasonic calls. Bat detectors should not be regarded as the sole identification tool but should be used to augment and amplify information gleaned by our other senses.

This book describes methods for identifying individual species of bats, how to find roosts, feeding habitats and commuting routes. We hope that not only bat workers but many others involved in conservation will also find it useful: ecologists, habitat assessors, environmental consultants and arboriculturalists.

A compact disc with recordings of the calls of British bats accompanies this book.

BB
DK

ACKNOWLEDGEMENTS

We would like to thank those friends and colleagues who read and commented on the early drafts of the manuscript, particularly Peter Geary of the Herts and Middlesex Bat Group and Sheila Wright of the Sussex Bat Group.

Thanks also to Colin Catto and Tony Hutson of The Bat Conservation Trust who provided expert comment on the final draft.

John Dobson of the Essex Bat Group gave us permission to use his recording of the Leisler's bat.

Thanks to Kees Kapteyn, Herman Limpens and Wim Bongers who led the first European bat detector workshop, and showed what was possible.

ABOUT BATS

- Bats are the only mammals with powered flight.
- There are about 970 species world-wide; almost a quarter of all mammal species.
- They are divided into the Megachiroptera (170 species) and Microchiroptera (800 species).
- The Megachiroptera are mainly fruit eaters and live in the Old World tropics.
- The Microchiroptera eat a variety of foods and, generally, navigate and forage using echolocation.
- Bats fly with their 'hands'. The bones of their fingers and forearms are elongated and adapted to support the wing membranes.
- The thumb is free of the wing membrane. Its claw aids climbing and grasping.
- The knee is rotated backwards and upwards so that the tail membrane can be lowered in flight and the bat can easily hang, head down.
- All 16 species of British bats are insectivorous Microchiroptera.
- There are two families of bats in Britain, the horseshoe bats and the vesper bats. They make up about a half our mammal population.
- Some species may live as long as 30 years.
- Our commonest bats are the pipistrelles which weigh about 5 grams and have a wingspan of about 200 mm.
- A pipistrelle may have a feeding territory with a radius of about 2 kilometres.
- Noctule bats may fly up to 6 km from their roost at speeds of up to 50 km/hr.
- Bats have good eyesight but use echolocation when light levels fall.
- To hear the sounds that bats use for navigation and foraging, we must convert their very high pitched sounds into the human hearing range with ultrasonic detectors (bat detectors).
- Bats produce many sounds that we can hear unaided by bat detectors. These sounds are mainly concerned with social interactions.
- Bats are nocturnal, becoming active around dusk and at dawn when the major night-time insect emergence occurs.

- British bat behaviour changes seasonally. Bats hibernate through the winter, arousing in the spring when the females begin to search for suitable roost sites to raise their single baby. The females gather together in the warm, summer months, forming maternity colonies. Babies are born in mid summer and are able to fly at about three weeks old. Fed on mothers milk, they are weaned by about six weeks.
- Lactating bats leave their babies in the roost when they fly out to feed at dusk and frequently feed again at dawn.
- Mating takes place primarily in the autumn. The males become territorial and will call females to their roost sites, defending them from other males.
- Some European bats migrate, travelling enormous distances across the continent to find suitable sites for hibernation.

RIGHT FROM THE BEGINNING

Most of us can recognise a robin, a blue tit or a blackbird, without a moments hesitation and we could probably add another dozen species to that list. Bird watchers, with relatively modest experience, will be able to identify a hundred or more different species and the experts will have lists of four or five hundred in Britain alone.

Bats are nocturnal mammals but they share, with birds, the ability to fly.

We can use many of the same clues that we use to identify common birds to identify bats in flight. So how do the birders do it?

Behaviour

Perhaps the most important aid to the identification of any animal is a thorough knowledge of its behaviour.
Where does it feed?
What does it feed upon?
Where does it roost?
In which part of the available habitat is it most likely to be found?
Does this behaviour change at different times of day?
How do the changing seasons affect its behaviour?
What is its mating behaviour?

You would not expect to look for a roosting swan perched in the branches of an old oak. You would not expect to see a heron chasing gnats in the evening. They are not designed for it. They are just not built that way.

Each species has evolved to exploit a particular niche, to fly in a particular manner, to take a particular type of food. The mallard can no more hover over a motorway verge than a kestrel can feed on weed from the bottom of a pond. But we could predict that just from inspecting the creatures, by looking at the shape of the wings, by relating the size of the wings to the size of the body and by examining the feet and beaks. So it is with bats.

A noctule bat could no more spiral inside the crown of an oak, gleaning insects from the leaves, than a long-eared bat could speed above the trees, diving for beetles over pasture. They are just not built for it. The shape of the wings and the size of the body determine the variety of foraging and flight behaviour and we can use these variations to predict where to find them and as clues for identification.

Sound

How many bird calls do all of us know? With practice, we can learn to recognise perhaps fifty and experts will make confident identifications from hundreds of different songs without having to see the bird itself, at all.

The use of detectors to make bats' **ultrasonic calls** audible to us has allowed us to enter a similar but hitherto secret world. Discovered by Donald Griffin as recently as 1938, the variety of calls used by bats as a group and as individuals is quite remarkable. Many of the sounds are species-specific. We can use these calls and their variations to identify the type of bat and study its behaviour.

These calls will be described in some detail later on.

Jizz

"Birders" know about **Jizz;** that almost indefinable something that characterises how a bird perches on a branch, how it holds itself, how it moves through the air. When an artist paints or sculpts a bird, we can instantly recognise if he or she has got it wrong.

We can describe the swooping, darting, harum-scarum, erratic flight of the pipistrelle and recognise it as different from the way a whiskered bat of similar size steadily patrols along the tops of hedgerows.

Geography
Where are we?
A large raptor soaring over Salisbury Plain is more likely to be a buzzard than an eagle. Serotines, horseshoe bats and grey long-eared have restricted ranges and the distribution of other species, such as Leisler's, is patchy.

Size
How big?
What size is the bird? Is it a pigeon or a wren?
With bats, is it a noctule or a pipistrelle?

Bats can be divided into three groups, by size -
Large: noctule, serotine, greater horseshoe, Leisler's.
Small: pipistrelle, whiskered, Brandt's, lesser horseshoe.
Medium: all the others.

Shape
Look at the silhouette.
Does the bat have long, thin wings like a noctule or broad wings like a serotine?
A noctule bat has a wedge-shaped tail. The ears of a long-eared bat are often easy to see.

Colour
Sadly, colour is of no great help.
A spotlight may pick out a pale underside, but beware. Most bats will look pale underneath if the light is bright enough.

Practice
Practice, practice! Nobody would expect to be able to tell a treecreeper and a short-toed treecreeper apart without many, many hours in the field.
It is no different with bats. To the echolocation calls we must add flight style, habitat usage, behavioural variations and other factors before we can make confident identification.

WHY DO THEY FLY LIKE THAT?

1) Why do different species of bats fly in different ways?

2) How do these different flight styles affect their hunting behaviour?

3) What consequences does this have for their general behaviour?

4) Can we use this information to identify individual species of bats?

Specific flight behaviour is related to the size of the bat, its weight and the shape of its wings. These factors determine the flight speed, manoeuvrability and agility, its ability to hover and the most effective flight style for the economical use of energy. All animals face a continual struggle to maintain a balance between energy expenditure and energy input. It is, simply, a balance between the costs of foraging and the energy gained from the food that it acquires.

All bat species have some competence in most flight patterns, but variations in body size, wing form and type of ultrasonic call determine which particular niche, in the habitat, is exploited by the bat. There are mathematical formulae that can be used as a sort of shorthand to describe some of these factors.

Wing loading (Defined as weight/wing area), describes the relationship between the wing size and the mass carried in flight by the bat. This mass will be made up of the weight of the body of the bat itself (which will of course vary considerably, not just through the year but throughout each twenty-four hour period), plus any food, embryo or young that it might be carrying.

Aspect Ratio (Defined as wingspan2/wing area), concerns the breadth of the wing. Fast flying bats with long narrow wings, such as the noctule, have high aspect ratios, whereas the slow flying brown long-eared's wing, which is broad, has a low aspect ratio.

Increasing the wing loading, together with an increase in the aspect ratio will give faster flight.

The **Angle of Attack** describes how much of the **leading edge** (the front of the wing) is lifted into the air stream, as the bat flies and is related to the stalling speed. Fast flyers have small angles of attack and have a higher stalling speed. Bats with broad wings, that fly more slowly, have maximal angles of attack and a greater ability to increase the **camber** (front to rear curvature) of their wings This results in a much greater ability to fly slowly or even hover.

WING SHAPE AND BEHAVIOUR

We can examine the wing shapes of different species of bat and predict how they will fly and forage.

The noctule has high wing loading and high aspect ratio; a heavy bat with narrow wings. The serotine of similar size and weight has broader wings (lower aspect ratio). This enables the serotine to fly and manoeuvre much closer to vegetation than the noctule which, because of its 'design', is most often found flying high and fast in the open, feeding over open pasture, rather than close to foliage. It would be likely to stall at the slower speeds that such manoeuvres, close to vegetation require.

Other predictions are also possible from an examination of these factors. For example, the noctule has long, narrow wings, high wing loading and high aspect ratio which requires fast flight.

(Figure 1)

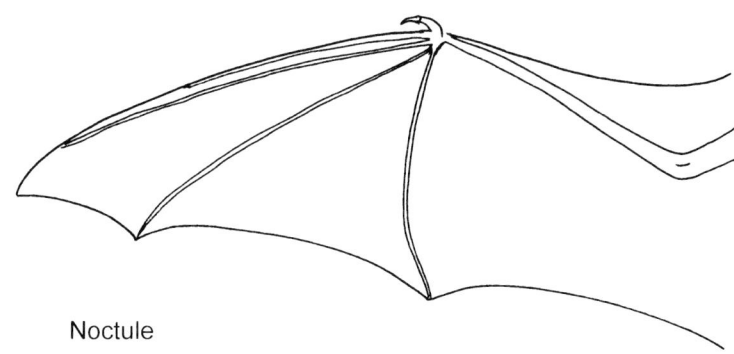

Noctule

This allows a rapid dispersal to distant hunting areas. Because the noctule is not very manoeuvrable in confined areas, it will need to hunt in the open. The consequence of its high stalling speed will tend to limit it to roost sites where it can free-fall, to launch from the roost entrance, with enough speed for flight. So we can predict, with some degree of certainty, that it will be found roosting high up in trees, at the edges of woods or in clearings.

On the other hand, bats with short, broad wings and low wing-loading (e.g. brown long-eared), have slow 'energetically expensive' flight and thus do not travel long distances to feed.

They will need good foraging areas, near to the roost and consequently, the numbers of bats in each roost will tend to be smaller. They will however be able to carry larger prey items, such as moths, to feeding perches. Add rounded wing tips and a large tail membrane to the original characteristics and we are describing a bat that can hover, and make tight turns in slow flight. It can glean from foliage.

(Figure 2)

Long eared

Different areas of the wing perform different functions.
The Arm wing (plagiopatagium) and Tail membrane (uropatagium) are primarily load bearing.
The Hand wing (dactylopatagium) produces the thrust for forward movement and for hovering . The Leading edge (propatagium), the wing membrane at the front of the arm bones, determines the angle of attack and hence the stalling speed.

(Figure 3)

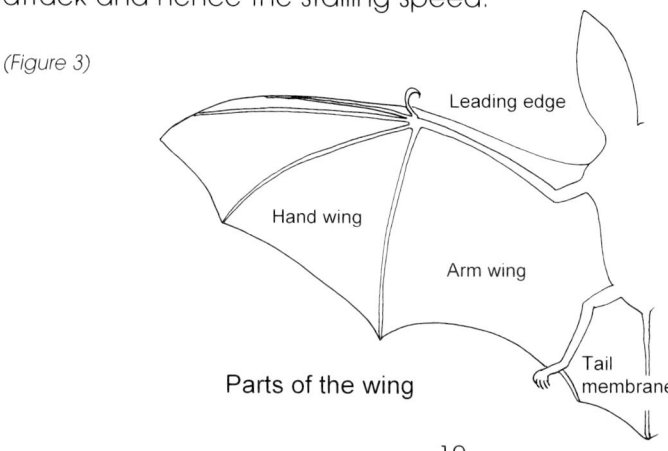

Parts of the wing

Both the Natterer's and the Daubenton's bat frequently feed over water. Because of the differences in the shape of their wings, they can forage in different ways and we can use these differences as clues to their identification. The area of the propatagium in the Natterer's is larger than the Daubenton's. This allows the Natterer's a much greater range of speed and a greater ability to make tight turns. The Daubenton's tend to fly in wide curves, parallel to the water surface, accelerating and decelerating, relatively smoothly. The Natterer's makes frequent changes of speed, often changing its angle of orientation to the water surface, coming up from the water to feed in the overhanging foliage and making frequent tight turns of up to one hundred and eighty degrees. The difference of speed changes can be heard in the changing rates of pulses in the bat's echolocation calls.

PHYSICS OF SOUND AND THE JARGON

You can go out into the night, switch on your bat detector and listen. Of course, you will hear bats and, with practice, will be able to make some identifications. But to explain what bats are doing and how they echolocate and why one call is like a 'click' and another like a 'smack' requires some understanding of how the sounds are constructed and of the terms that describe the various differences.

How do we explain:

that a noctule bat's call sounds different as it flies lower?

that you can hear the same horseshoe bat at different frequencies?

that pipistrelles and whiskered bats sound alike at 60 kHz?

that '45' pipistrelles produce 'clicks' when '55' pipistrelles produce 'smacks', when you listen at 55 kHz?

Wavelength and Frequency

This section will try to make some sense of the technical terms in common use. The following section, '**Describing Ultrasound**' will give some of the theoretical background to the practical skills of bat detecting explained in the section '**Interpreting the sound of the Heterodyne Detector**'.

Either read on now or skip to the practical section and review the theory later.

Sound is propagated as compressions and rarefactions of the medium through which it is travelling (in this case air), spreading out spherically, in three dimensions, from the point of origin. The ripples from a stone thrown into a pond give a two dimensional impression of the process. Sounds are usually represented, diagrammatically, as lateral waves.

Wavelength is a measurement of the distance between successive peaks and troughs of the sound wave. Wavelength is, of course, related to the frequency. As the frequency increases, the wavelength shortens.

Frequency describes the number of wavelengths per second.

One sound wave per second is called 1 Hertz, named after Heinrich Hertz the German physicist. Because sound travels in air at 340 metres per second, 1 cycle per second (1 Hz) has a wavelength of 340 metres.

(Figure 4)

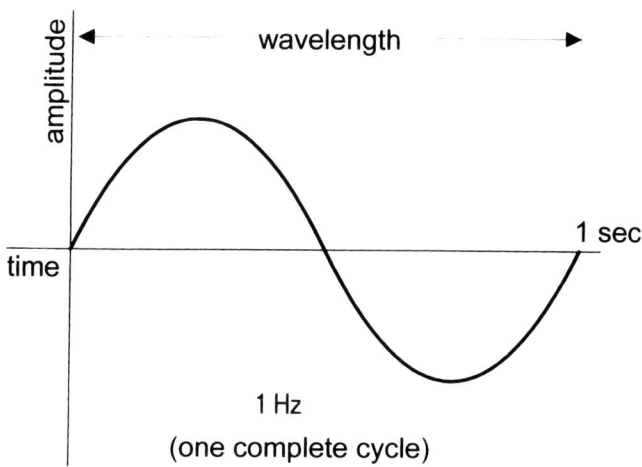

So, at 10 cycles per second (10Hz), the wavelength is 34 metres.

(Figure 5)

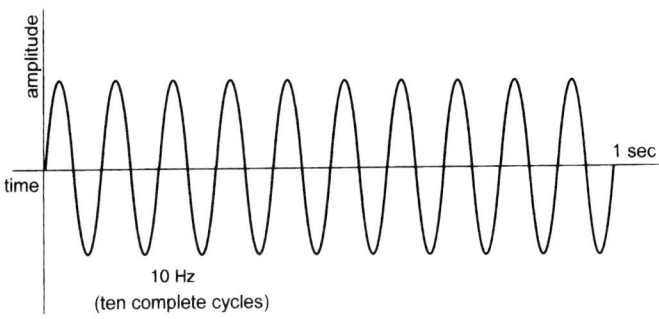

1000 cycles per second is called 1 kilohertz (1 kHz). These are the units that are used when we talk about bat sounds.

At 50 thousand cycles per second (50 kHz), a frequency used by most of our bats, the wave is only 6.8 mm long, and gives a distinguishable echo, virtually the same size as the insect prey that the bat is hunting. These echoes provide very good information about the prey. This is known as **TARGET DISCRIMINATION.**

Humans can hear frequencies up to about 20 kHz. All frequencies above this are called **ULTRASONIC.** Dogs' and cats' hearing extends into the ultrasonic range.

Sound waves with long wavelengths behave differently from those with short wavelengths. Long waves can travel large distances and bend round objects in their path, rather than being reflected back from them as echoes, so such long waves are of no use to a bat.

Very high frequency sounds do not travel very far in air unless they are extremely loud, i.e. they are rapidly **ATTENUATED.**

The **AMPLITUDE** of a sound wave is a measurement of the loudness and determines the amount of energy that the wave contains.

Harmonics

Nearly all sounds are made up of a mixture of frequencies. In the case of musical instruments, this complex of sounds enables us to tell a violin from a trumpet or a flute from a clarinet and is mostly due to the fact that when played the different parts of the instrument all vibrate together, producing sound waves of different frequencies.

These different frequencies, in a complex sound, are called **HARMONICS.** It is not a random collection of noises, but a precisely ordered sequence of related frequencies. The lowest or FIRST HARMONIC is called the **FUNDAMENTAL.** The harmonics are exact multiples of the fundamental frequency i.e. x2, x3, x4, and usually are produced with much less energy than the fundamental.

A single note that does not vary in pitch is said to be a **CONSTANT FREQUENCY (CF). (Track 1)**

A sound that varies in pitch is said to be **FREQUENCY MODULATED**. This can be from high to low or vice versa. Most bats that use this type of call modulate their frequency from high to low. **(Tracks 2 and 3).**

Some bats use calls that frequency modulate, others use constant frequency and some use bits of both. If the bat modulates (changes) in frequency fast, at the start of the call, and then gradually slows down the rate of change it will be varying from **STEEP FM to SHALLOW FM.**

Some bats begin their call with a **FREQUENCY MODULATION** and end it with a near **CONSTANT FREQUENCY** 'tail'; typical of our commonest bat, the pipistrelle. **(Track 4)**.

The CF call of the horseshoe bat begins and ends with a short FM sweep.

(Please note: Tracks 1, 2, 3 and 4 are synthesised examples and not real bat calls.)

Full details of all the tracks on the CD mentioned in the text will be found in the section "Index of CD tracks and commentary" on page number 53

DESCRIBING BAT ULTRASOUND

How do we talk about ultrasound, and how do we describe the noise of a bat to somebody else? How would you describe a colour?

The Oxford Dictionary defines it as:

"The sensation produced on the eye by rays of light when resolved by a prism or selective reflections into different wavelengths"

Alternatively, colour could be described in terms of "Corn fields, sunsets and roses" etc.

In the section dealing with individual species identification, there are some rather fanciful descriptions of bat sounds as heard through a heterodyne detector; 'thigh-slapping', 'stubble-burning', 'tap-dancing', 'hair-rubbing'. These convey a very subjective impression of how bats sound, relating to sounds with which we are already familiar, and we can each invent our own analogies. But first, let's get an idea of how to make a rather more formal description.

Sonograms
A sonogram is a graphical representation of a bat's call. The vertical axis shows the frequency range. The horizontal axis shows the length of time taken by the call.

What factors are most useful for precise description of ultrasound?

Length of Call
Individual calls can vary from about 0.2 to 100 milliseconds (100 msec). One millisecond = one thousandth of a second.

Harmonic structure
Using a number of harmonics increases the range of frequencies **(BANDWIDTH)** of the call. Some bats, such as the greater horseshoe, will suppress the fundamental (first harmonic) and put most of the sound energy into the second harmonic. The fundamental can however be heard in certain circumstances. Other bats such as the brown long-eared expand their basic call with a number of harmonics. One pulse

will contain a 'stack' of harmonic FM sweeps, getting higher and higher, each one slightly overlapping the one before. This gives them the much more detailed information that they need, in order to navigate within the foliage.

Frequency modulation (FM) (Tracks 2 and 3)

FM is typical of Myotis bats, which use a range of frequencies in each call. The extent of the range and the speed with which the call sweeps down, through the different frequencies, will vary. When seen on a **SONOGRAM,** it might appear as a near-vertical, linear sweep, sometimes slightly flatter at the lowest frequencies. Vertical sweeps are of very short duration. These are typical calls of most small, insectivorous bats and give considerable information about the texture of the surroundings (e.g. insects resting on leaves) and allow accurate ranging of the target. *Bat pulses will be hundreds of times faster than the artificially generated examples on these two tracks and too fast for us to perceive them as frequency changes. They are heard as 'clicks' on a 'heterodyne' bat detector. With 'time-expansion' detectors it is possible to slow the pulses down enough to hear them as frequency sweeps. (Figure 6)*

Constant frequency (CF) (Track 1)

These pulses are usually of longer duration and the frequency used is normally species-specific. Constant frequency pulses are typical of the horseshoe bats, the greater horseshoe calls at around 80 kHz, the lesser horseshoe at 110 kHz.

(Figure 6)

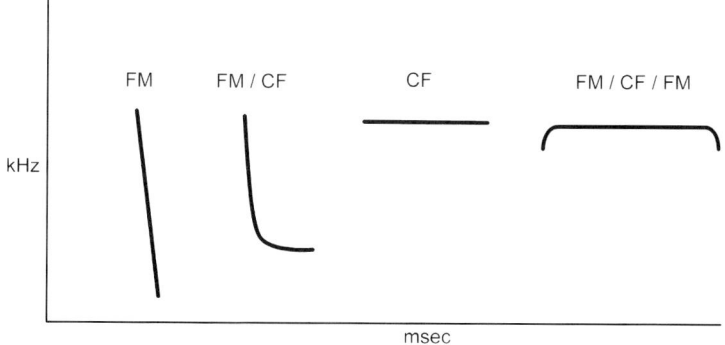

When the pitch of a sound rises as the sound source approaches and drops as it goes away, this is known as the DOPPLER SHIFT. It is particularly important to those bats whose echolocation is primarily of constant frequency. The echo received from a flying insect will be higher in pitch than the original echolocation call when the insect is approaching and lower when the insect is flying away. Horseshoe bats use this information to judge target range and they can alter the frequency of their call so that the returning echoes stay at the frequencies to which their hearing is most sensitive.

CF and FM

Some calls are a mixture of both CF and FM. Horseshoe bats have short FM components at the start and finish of their call (FM/CF/FM). These are important in judging position and distance whilst the CF signal enables the bat to detect and distinguish the wing beats of flying insects.
(See figure 6)

Shallow FM/CF (Track 4)

Starting with an FM portion, the call changes frequency much more slowly as it reaches the lowest frequencies. The latter portion of the pulse is called the 'tail' which, on a sonogram, looks virtually horizontal. This is typical of bats hunting away from clutter e.g. noctule. (Figure 6)

Amplitude

Calls vary in amplitude from one species of bat to another. Generally, brown long-eared bats produce very quiet echolocation calls whilst noctules are very loud. There is also variation of the amplitude in different parts of a single pulse. The whiskered bat uses an FM call that sweeps from around 70 kHz down to around 30 kHz, but it is loudest at about 45 kHz.

The perceived amplitude will also depend upon the distance of the bat from the observer. The higher frequencies at the top of an FM sweep will be more attenuated than those at the lower end and this will alter the sound heard on the detector, giving a duller sound. Thus a Natterer's bat, heard from a distance, will lose the 'crispness' of its call and sound more like a whiskered bat.

VARIATIONS IN A CALL

Each bat's echolocation call will change as it flies from one type of habitat to another and as it locates **(search phase)**, chases **(approach phase)** and catches **(terminal phase)** prey.

Bats can vary their foraging strategies in different habitats so, in order to obtain maximum information about their surroundings, they must change their echolocation calls.

Bats that primarily fly in **OPEN SPACES** will :
- use lower pulse repetition rates.
- use echolocation that is long range, loud and more regular..
- have a longer constant frequency component.
- change direction less frequently.

Bats that fly in **CLUTTER** on the other hand:
- use faster pulse repetition rates.
- have calls which are shorter and quieter.
- use frequency modulated calls (steep FM).

When bats of different species fly in similar habitats they will require similar information about their surroundings and so their echolocation calls will also become similar.

A noctule flying high above the trees uses a pulse 15-20 msec long. **(Track 5)** At street lamps, this is shortened to 12-15 msec and in a clearing in a wood it will be as brief as 3-6 msec. Serotine bats feeding at street lamps have a pulse length of 13-14 msec and Myotis bats, in woodland, use pulses 3-6 msec long. (see below).

As the noctule descends and comes closer to the vegetation or within a woodland ride it will drop the CF portion of its call, because it requires different information about its environment. **(Track 6)**

AVERAGE PULSE DURATION (milliseconds)

	HIGHER THAN 30m	STREET LAMP	CLUTTER
NOCTULE	15-20 msec	12-15 msec	3-6 msec
SEROTINE		13-14 msec	
MYOTIS			3-6 msec

This can cause confusion because seemingly, when we listen to this on a heterodyne detector, a noctule flies in, what sounds like a 'serotine' feeds and we get the impression of a 'whiskered' flying off !

Echolocation and insect capture

Insect capture can be divided into three distinct phases. **SEARCH, APPROACH and TERMINAL**.

(Figure 7)

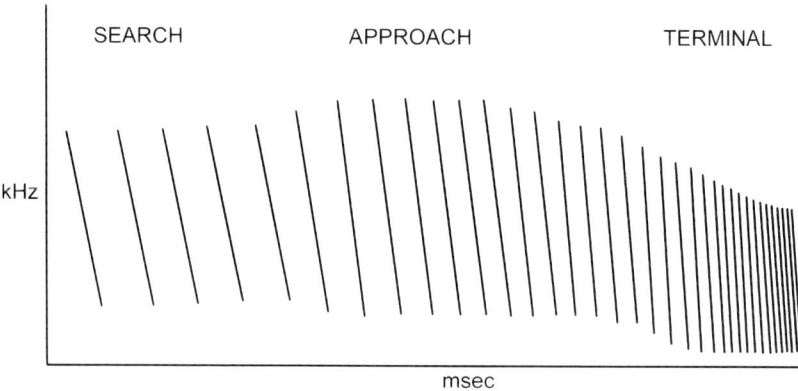

This generalised sequence of events seems to be common to all bats and each phase is marked by a change in the rate of emission of pulses and the duration of each pulse.

There is also a change in the harmonic structure of each pulse and in the amount of CF or FM signal in each call.

The calls are loudest during search phase and progressively quieten until insect capture is attempted.

Pulse repetition rate increases as the bat locates and then attempts to capture the prey. During the search phase, pulses are emitted at about 10 per second. This increases to 20-40 pulses per second once an insect target has been identified (approach phase) and may reach 200 pulses per second in the moments before the insect is caught. (terminal phase). **(Tracks 9 and 10)**

In CF bats, the duration of the signal will be shortened and the FM component of the call increased. Bats, such as serotines, that have a short CF tail to their primarily FM signal, will shorten the call and leave out the CF tail, as they move into approach phase.

During the approach phase the FM sweep of the harmonics may be increased. In the terminal phase, the overall FM sweep is lowered slightly and the bandwidth of each harmonic is reduced. This would only be noticeable by detailed computer analysis.

The terminal phase is heard on a heterodyne detector as a buzz. This '**feeding buzz**' often drops into the upper range of human hearing. **(Tracks 7 and 8)**

METHODS OF DETECTING ULTRASONIC CALLS

There are three methods of converting ultrasound into audible sounds, each with specific uses, advantages and disadvantages.

This book concentrates on the use of the **HETERODYNE DETECTOR.**

With a Heterodyne detector the bat's ultrasonic call is converted into an electronic signal which is compared with a variable internal signal, generated by the detector, and we listen to the difference between these two signals.

The internal signal is varied by turning the tuning dial which allows us to listen to a limited range of frequencies, usually about 10 kHz.

The Heterodyne detector gives high sensitivity, but with a limited frequency bandwidth. However, it emphasises call differences and enables us to determine the range of frequencies in the field. The limited bandwidth makes it hard to search over the entire sweep of the call, requiring manual tuning and there is no frequency information in any recordings.

Very expensive models have a scanning heterodyne system that searches over a range of predetermined frequencies, automatically, and is a very useful system for field work.

Frequency Division

This system transforms the whole ultrasonic signal into audible sound without tuning. The incoming signal is checked by counting the 'zero crossings' of the sound pressure waves and produces a single wave for every predetermined number of incoming waves, e.g. 1 pulse out for every 8 or 10 incoming pulses, which consequently lowers the frequency into the audible range.

Frequency division has the advantage that we can listen in real time to the whole of the bat's call and all the amplitude and frequency information is retained in recordings. It is however not a pleasant noise to listen to and the harmonics are not retained. It is more useful for recording and subsequent analysis of the sounds by computer.

Time Expansion with Digital Memory

Time expansion systems store the whole of the incoming signal in a digital memory and then play it back at a slower speed, either immediately or later in the laboratory, for computer analysis.

The signals retain their true proportions. When the frequency is lowered, say, 10 times, the signal is also extended in time by a factor of 10. There is no loss of information, including harmonics, and the system is broadband. However storage time is limited by the memory capacity and recordings are not in real time. Using time expansion, pitch and pulse shapes are easy to hear.

The use of time expansion and frequency division, as adjuncts to heterodyning, has the advantage that any bat sounds that may not be detected by the limited frequency range of the heterodyne detector will still be heard using these two broadband systems. Time expansion facilities should be used as an addition to heterodyning, which remains the most important method of ultrasound detection in the field.

INTERPRETING THE SOUNDS OF THE HETERODYNE DETECTOR

How can we make sense of all the various ultrasonic noises that we hear?

We have to remember that the heterodyne detector is a very selective tool, with a bandwidth of about 10 kHz. Because the frequency range of most bats is greater than the bandwidth of human hearing, for most bats we will only be hearing a small portion of each individual call. Turning the frequency knob allows us to choose which part of the call we hear.

All bat identification begins with learning to differentiate between two types of call. Those that sound like a 'dry' **click, (Track 11)** from those that sound like a 'wet' **smack, (Track 12).**

Very short sounds, heard on the detector sound like 'dry' clicks and have no real tone. As the sounds become longer, they acquire more tonal quality (the 'wet smack'), as in the 'boing' of the tympani, compared to the short, sharp 'rat-a-tat-tat' of the snare drum.

A single pulse of bat sound may be composed of both fast and slower elements which can be tuned into separately and so we are able to hear both 'dry' and 'wet' components in a single call. Some bats will only make very short, sharp bursts of sound which will be heard as 'dry' clicks over the whole of their frequency range.

This basic listening skill is the foundation for using a bat detector.

First steps in identification:
We could try to picture the sonogram as we turn the dial. Imagine tuning the detector up and down the curve, making sense of the changes in sound that we hear.
Is it a vertical sweep?
Does it flatten out towards the end and hence change the sound quality?
Is it a single note throughout (as in the call of the horseshoe bats)?

Bats that just use short, sharp calls that almost instantly descend through a whole range of frequencies are known as **frequency modulating (FM) bats.** On the detector, they are heard as **'dry clicks'** throughout the tuning range. Typical of these is the Daubenton's bat. **(Track 36)**

Other bats begin their call with a fast downward sweep that sounds like a 'dry' click, but they slow down the last part of the call so that the end sounds like a 'wet' smack. Typical of these is the pipistrelle. **(Track 15)**

We can often find pipistrelles and Daubenton's flying together, over water. Practice listening to both types of call, with the detector set at around 45 kHz.

Tune the detector up and down, and notice that when you listen to the Daubenton's you hear clicks over the whole range, but when you listen to the pipistrelle, as you turn the frequency dial downwards the sound changes from a 'dry' click to a 'wet' smack. Notice the frequency at which this change in tonal quality occurs. **(Tracks 13 and 14)** This is the echolocating frequency which gives its name to the two kinds of pipistrelle bats.

Give yourself the best chance of making an identification. All bats change their echolocation calls as they fly and forage. Bats that are feeding or just leaving the roost, although easier to detect, will often not be making the calls that are best for identification purposes. It is possible to identify, fairly confidently some species of bats in many different situations.

Try to listen to the bat when it is flying from the roost site to the foraging area, before it starts to feed, or when it is returning to the roost later on. Listen when it is using search phase echolocation calls which are much more species specific.

Step back twenty or thirty yards from a feeding area and listen to the bats as they fly in, rather than as they are feeding.

Start your search with the detector on medium volume and set at 45 kHz. Most bats' calls sweep through this frequency.

What should you be taking note of, in order to make identifications?

1) What is the best listening frequency?
2) Are you listening to 'dry' clicks or 'wet' smacks?
3) What is the rhythm of the call?
4) How loud is the bat?
5) Is there a point where the clicks turn to 'wet' smacks?
6) At what frequency does this happen?
7) How fast are the pulses coming?
8) What is the frequency range?
9) Is there the same amount of energy throughout the whole of the call?

USE HEADPHONES

To attend to all these points requires concentration, practice and the best sound quality your machine can give you. Listening though headphones will not only give you better sound but will also increase the audible sound bandwidth and prevent you from being distracted by noises around you, particularly those from other bat detectors.

Different makes of detector have appreciably different tones and sounds. Once you have chosen the detector that suits you best, stick to it. Learn to recognise the sounds from that machine.

Different models of bat detectors have different bandwidths. This means that different ranges of frequencies are heard either side of the frequency at which the detector is set. If the bandwidth is +/- 8 KHz, it means that when your detector is tuned to 40 kHz for example, you will hear frequencies from 32-48 kHz.

THE ZERO POINT

Earlier sections of the book have described the characteristics that we should listen to when trying to make identifications using heterodyne bat detectors. We have learnt to recognise the loudness of the call, the rhythm, the repetition rate of pulses, the range of frequencies used in each call and the presence of 'wet' smack or 'dry' click. We can also listen for the lowest emitted frequency in each call; the end point.

Bats like the pipistrelle, serotine and noctule start their calls with a steep, fast, near-vertical sweep downwards, through a range of frequencies (FM sweep) and then hold on to the last frequency. If you imagine the shape of the call, (the sonogram), it looks like a 'J' written backwards. The bottom of the 'J' represents the lowest part of the call, called the **end point.**

It is the difference in frequency of this end point that enables us to separate the two phonic types of pipistrelle; the '45' from the '55'.

But how can we do this accurately?

Heterodyne detectors work by comparing an internally generated signal with one produced by the bat, and we change the frequency of this signal when we turn the tuning dial. The electronics within the detector allow us to mix the two signals and give the difference between them at the output (speaker or headphones).

The whole point of the exercise is to transform the bat's ultrasonic signal into a sound that we can hear. If our detector is tuned to 40 kHz and the bat is echolocating at 45 kHz, and we listen to the difference between these two frequencies, we hear a sound at 5 kHz, well within our hearing range which is from 20Hz to 20 kHz.

If we had a detector and we listened to the difference, when its pure signal (45 kHz) was tuned exactly to that of the 'perfect' bat (45 kHz), the result would be 0 kHz difference. NO sound. But because the bat's call is composed of a range of harmonic frequencies we always hear something.

This theoretical point of no sound or 'least difference' is called the **ZERO POINT.**

What we have to do in practice, when we search for the 'tail' of the pipistrelle's call, is to find this zero point.
The end point of the bat's call is 45 kHz, we have tuned the detector to 40 kHz, giving a difference of 5 kHz which is what we are listening to.

(Figure 8).

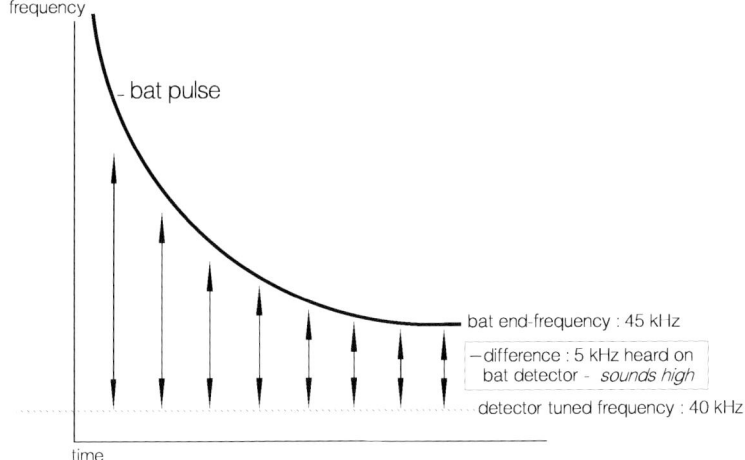

As we tune the detector upwards, the difference between the tuned frequency and that of the bat becomes smaller, so that when our detector is tuned to 46 kHz the difference is only 1 kHz.

(Figure 9).

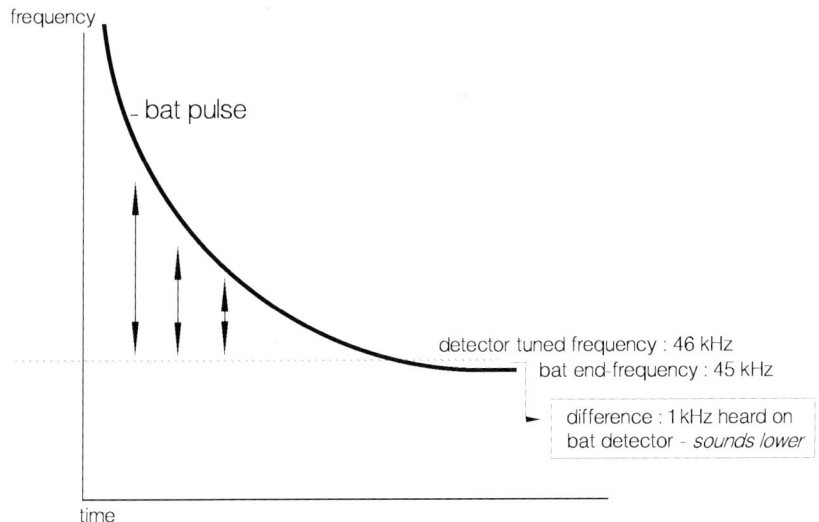

Continue tuning upwards, there comes a point when the tuning dial reads 50 kHz. The difference between that and the bat's frequency is now 5 kHz.

(Figure 10),

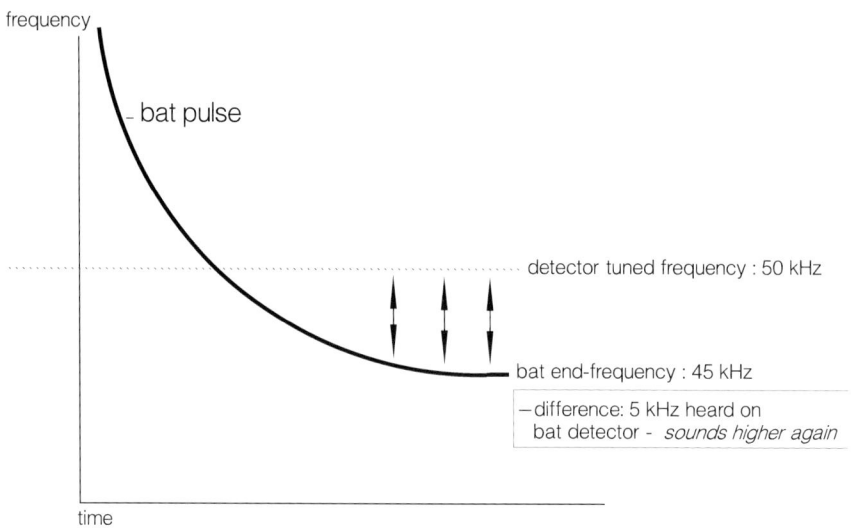

5 kHz is a higher pitched sound than 1 kHz.

We know that the 'flat' tail of the pipistrelle's call sounds like a 'wet' smack.

To find the end point therefore, keep turning the dial up and down, trying to find the point at which the 'wet' smack sounds lowest. This is the nearest we can get to the zero point.

As we listen to the vertical FM part of the call we hear 'dry' clicks. Further tuning downwards produces firstly high pitched 'wet' smacks, which will become lower pitched 'wet' smacks as we find the 'tail' and then rise in pitch again, if we continue to tune beyond the tail.

The reverse is that, if we start listening below the bat's tail frequency, as we tune the detector upwards, towards the zero point, the pitch of the 'wet' smack will go down as we reach the zero point and then rise again as we tune up above it.

The position of the lowest pitch, on the tuning dial, will separate the two phonic types.
(Track 46)

Also, listen to **Tracks 41** and **42** to hear the change in pitch when the tuning for greater horseshoes is offset by 5 kHz. **Track 21** shows the effect of offset tuning on the pipistrelle's call and **Track 22** has the FM component of one pipistrelle bat at the same time as the near CF tail of another pipistrelle flying with it.

INDIVIDUAL SPECIES

Pipistrelle (Pipistrellus pipistrellus)

Pipistrelles are by far the commonest species, their flight is fast, darting and erratic and although like other small species they will follow linear features when they commute from roost to feeding area and they will be continually stopping to feed along the way. Typically, flying at head height, they will often feed higher up, early in the evening and come lower down as the temperature drops.

When crossing open spaces, they fly faster and increase the pulse repetition rate.

The echolocation pulse consists of a short, FM sweep ending in a longer, near-CF tail. By tuning the detector, the FM component of the call will be heard as a 'dry' click and, tuning the detector downwards, the CF tail produces a 'wet' smack. These are two parts of the same single echolocation call and not separate, alternating pulses, as used by noctule bats. **(Tracks 15, 16, 17, 18, 19 and 20)**

Now that two separate species of pipistrelle have been identified, with significant differences in their echolocation, it is important to note the position of the change of sound from 'dry' 'to 'wet', i.e. the point at which the CF tail is heard. **(Tracks 21 and 22)**

One variety, the '55' pipistrelle, has a near CF tail at about 55 kHz, whilst the other, the '45' pipistrelle has its CF tail at about 45 kHz. So, with the detector set at 55 kHz, you will be able to hear the CF tail of one species but will still be listening to the FM sweep of the other, whose CF tail will only become audible at 45 kHz.

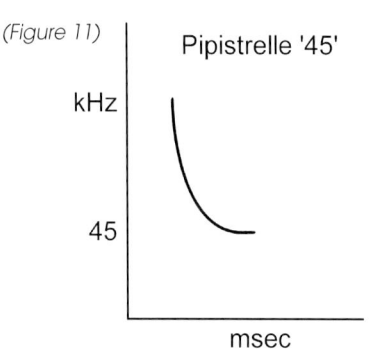
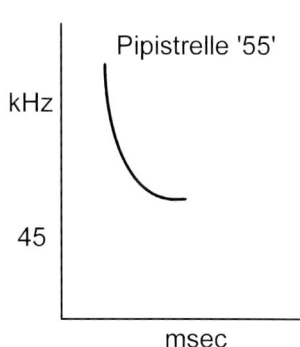

(Figure 11)

Another method of trying to separate one from another is by noting at what frequency the 'wet' sound disappears. If your detector has a bandwidth of 8 kHz then you should be able to tune downwards from 55 kHz and the 'wet' sound will almost disappear at around 47 kHz. Similarly, for the '45', the sound should vanish at about 38 kHz.

A more reliable method however, is to search for the **ZERO POINT**. This was explained in an earlier section, and is a critical identification feature for those bats whose calls end in a near-CF tail.

The distinction between '45' and '55' pipistrelles is not absolute. The majority of calls of '45's lie between 44 and 48 kHz and '55's between 52 and 56 kHz. For identification purposes, a call with a zero point below 48 kHz can be recorded as a '45' pipistrelle and a call with a zero point above 52 kHz can be recorded as a '55' pipistrelle.

If the zero point is found between 48 and 52 kHz, it will not be possible to assign the bat to either species.

Limited observations suggest that when both species are feeding in the same woodland, '55's' are to be found along rides and paths, within the wood and '45's' at the periphery. Although their diets are largely similar, '55' pipistrelles tend to eat more aquatic insects, and thus, may have a smaller habitat range than '45's.

When flying in groups, the position of the CF component in the call of individual bats can vary by up to 14 kHz.

The rhythm of the pipistrelle echolocation call is irregular and has been likened to 'slapping a naked thigh with a cupped hand'!

Pipistrelles roost mostly in buildings. They are regularly found at woodland edge, parks, street lights and water. Social calls are commonly heard throughout the year. Loud 'chonks' at 20 kHz can be heard (often without a detector) which can, sometimes, be confused with the sound of a noctule bat. **(Tracks 23, 24, 25 and 26)**

Serotine (Eptesicus serotinus)   *(Figure 12)*

This is a large broad-winged bat that flies much more slowly than the similar sized noctule and forages much closer to vegetation. Serotines are usually found roosting in buildings and, true to their scientific name, they emerge later. They are often found over park land or cut grass, flying at about roof height.

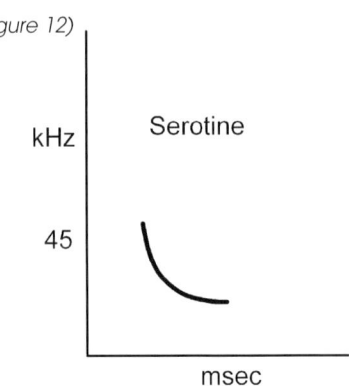

The call is very distinctive, with an irregular rhythm that has been likened to 'a poor jazz drummer' or 'clog dancer'. They make a loud 'tock' which is best heard at around 27 kHz. **(Tracks 29 and 30)**

Without having the tone variety of a noctule, the serotine's pulses tend to be woollier and duller. You can sometimes pick up the second harmonic of the call at 50-60 kHz. The sound quality will be different, but the rhythm unmistakable. **(Tracks 9, 10 and 12)**

Serotines, flying along hedgerows or commuting, may sound like Myotis myotis (the greater mouse-eared bat).

Noctule (Nyctalus noctula)   *(Figure 13)*

The noctule's long, narrow wings and high wing loading give it fast flight, often high above the trees. Typically, it hunts with steep dives over open pasture, to take larger insects. The noctule's loud, long range calls can be heard up to 200 metres away. It has a slow rhythm in the open, sounding like 'chip-chop'; an FM sweep followed by a second pulse that is near-CF. **(Track 5).**

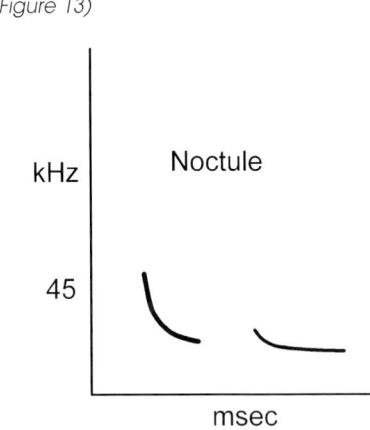

The best listening frequency is around 20 kHz, and harmonics may be picked up at 40 and 60 kHz.
As the bat flies lower, below tree top height or in woodland glades, the CF component of the call is omitted and the pulse rate increases. **(Track 6)**

Noctules are among the first bats to emerge in the evening, often before sunset. They are easy to see, and the wedge-shaped tail can be noticed. They will nearly always be found roosting in trees.
(Listen to Track 31 which has noctule and serotine bats flying together.)

Leisler's Bat (Nyctalus leisleri)

The Leisler's bat is smaller than the noctule, its flight and hunting style is similar but less dramatic. The dives for prey are shallower. They are more likely than noctules to use buildings as roost sites. Leisler's bats usually emerge when the noctules have returned to their roosts, after their initial feeding spell.

The Leisler's echolocation call is similar to the noctule's, but they use alternating FM and CF pulses less often. It is best heard at around 25 kHz. If the detector is tuned upwards, dry clicks will be heard. **(Track 32)**

(Figure 14)

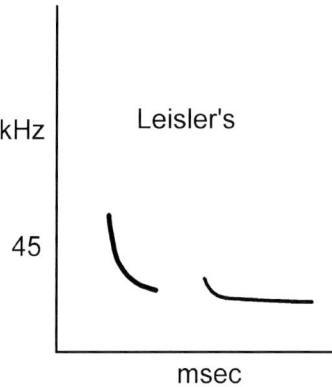

Like the noctule, Leisler's drops the CF portion of the call as it flies lower and closer to 'clutter', and the pulse repetition rate increases.

Both Leisler's and noctules may omit the FM part of the call when flying high, and the frequency of the remaining CF pulse can vary. They tend to lower the frequency of the CF pulse as they fly higher into an increasingly uncluttered environment.

Long-eared bats (Plecotus auritus and Plecotus austriacus)

Long-eared bats can hear the noises of insects on foliage and may 'switch off' their echolocation altogether, on occasions, whilst foraging.

The brown long-eared bat's echolocation is very quiet. Some detectors are not sensitive enough to pick it up at all. Most require the listener to be within a few feet of the bat. The bat may catch the eye, as it forages around a tree, but often nothing is heard on the detector at any frequency setting.

(Figure 15)

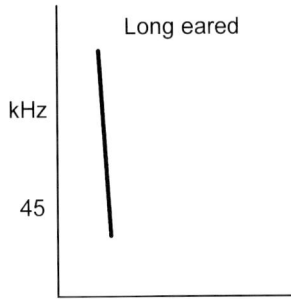

Because it is feeding within or very close to the foliage, it uses steep FM sweeps with a rapid repetition rate and a number of harmonics. These bats make sounds like the ticks of a Geiger counter and are usually best heard at around 50 kHz. **(Tracks 27 and 28)**

Typically, they fly slowly up and down the foliage of a single tree, a number of times, before moving to their next feeding site. Long-eared bats will hover to take insects from the leaves. They will also be found flying around bushes and have been noticed, on occasions, to fly from the roost to large enclosed spaces, such as barns, where a number will fly for a while before leaving to feed. When flying across open spaces the calls of long eared bats are much louder.
Brown long-eared bats make social calls that sound like 'treep' on a heterodyne detector when set at about 20 kHz.
Most people feel unable to differentiate brown from grey long-eared bats, by their echolocation calls.

Nathusius's Pipistrelle (Pipistrellus nathusii)

Nathusius's pipistrelle is best heard with the bat detector tuned to about 39 kHz. The pulses are longer than those of the common pipistrelles. On a heterodyne detector, longer pulses, at the same amplitude, sound 'heavier'. The common pipistrelle's pulse is about 5 msec long and sounds like 'plip-plip' and the Nathusius's with a pulse of 8 msec. sounds more like 'plop-plop'.

The first two breeding roosts of Nathusius' pipistrelle in the UK were found in 1997. It is worth paying particular attention to any bat with a near-CF ending to its call that has a zero point below 41 kHz.

Barbastelle (Barbastella barbastellus)

This rare bat has a call that is described as short, hard, with castanet like 'smacks'. It is best heard at about 32 kHz. The call begins with a CF component and ends with a short downward FM sweep. It is typified by intervals of fast and slow pulses. **(Track 33)**

The barbastelle flies slowly and may give the impression of pausing as it flies, rising a little from its level of flight, almost hovering, before it carries on.

MYOTIS BATS

The five species of Myotis bats found in the UK are the hardest to separate, one from another, in the field. All Myotis bats in Britain use fast, near-vertical sweeps of FM sound without any 'smacks' and hence, there is very little tonal quality in their echolocation calls which are heard as 'dry' clicks. (Track 11 - Brandt's bat compared to **Track 34** - pipistrelle and Myotis bat flying together.)

Because of these difficulties, we have to concentrate on other sound clues:

- Is the call loud or quiet?
- Is the repetition rate of the pulses fast or slow?
- How does the repetition rate vary?
- At which frequency does the call have maximum energy?

We also have to pay significant attention to the behaviour of the bat; its flight style, foraging behaviour and habitat usage.

Whiskered and Brandt's bats (Myotis mystacinus and brandtii)

It is probably not possible to differentiate these two species by echolocation alone, indeed it may not be possible to distinguish them in the field at all.

Flying along a forest ride, for example, they have a very regular rhythm of very dry clicks. The sound is louder and less sharp than that of a Natterer's, and the repetition rate is slower than either a Natterer's or a Daubenton's bat. Their sound seems to have a 'purposeful' quality. **(Tracks 11 and 35)**

With the detector set at around 55 kHz, these small bats can be differentiated from similar sized pipistrelles by tuning the detector downwards. The whiskered and Brandt's calls will be heard as 'dry' clicks throughout the range of frequencies, whereas the calls of the pipistrelle will become 'wet' smacks, as the detector is tuned downwards.

(Figure 16)

The intensity of the call is not constant, throughout the frequency range, but is strongest at around 45 kHz.

In flight, a whiskered bat may pass overhead and, if you wait a minute or two, it will come flying back again in the opposite direction. When feeding they have a settled flight in a horizontal plane, back and forth, along the top of a hedge, for example. When they have exhausted the feeding at a site, they will move to a new foraging area and settle again into a level flight. In flight, they have been likened to peacock butterflies.

Ingemar Ahlen reports[1]. that Brandt's bats are more likely to be found in forest or wooded park land but that whiskered bats are more frequently found in open habitat.

Daubenton's (Myotis daubentonii)

Typically associated with water, the Daubenton's bat hunts over lakes, streams and ponds. They generally fly within a few inches of the water surface, gaffing insects with their feet. They will come up from the water surface to the foliage, overhanging the water, and even fly to forage in adjacent woodland.

Daubenton's will also hunt in deciduous woodland. The flight style there is level and meandering at about 15 to 20 feet above the ground. They make strong, fast, 'dry' clicks, best heard at around 45 kHz, and they fly parallel to the water surface, making relatively wide turns. Whilst turning, the pulse repetition rate increases, with a gradual acceleration and deceleration, rather like a motor bike, and there is often a slight kink in the descending sweep of the sonogram. Under good conditions on a heterodyne detector, this can be heard as a 'two-stroke' call, "tik-ke-tik-ke". **(Tracks 36, 13 and 14)**

(Figure 17)

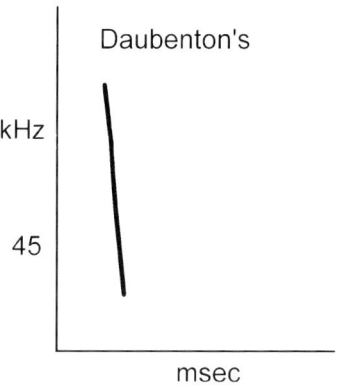

1. Ahlen, I. (1990). Identification of bats in flight. Swedish Society for Conservation of Nature. Stockholm

Natterer's bat (Myotis nattereri)

Natterer's bats are generally much quieter than the other Myotis bats. They specialise in hunting in confined spaces, making very narrow turns. They will emerge from the roost and often fly quite low, seeming to disappear into hedges. Their echolocation calls are very short with an energy peak at about 50 kHz. Often the repetition rate is so fast that one pulse seems to run into another, giving an impression of the fine crackle of burning stubble or the sound of hair being rubbed between finger and thumb, next to the ear. **(Tracks 37 and 38)**

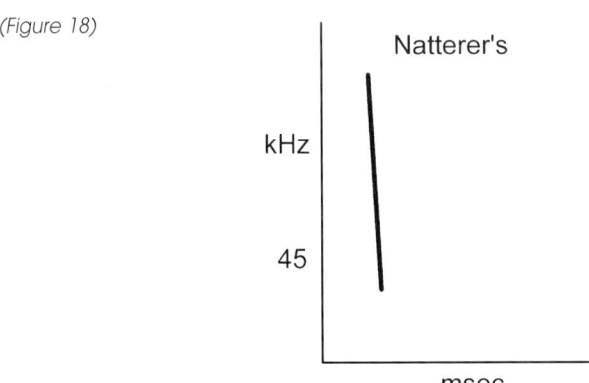

(Figure 18)

Natterer's bats are inquisitive and may fly up to an observer, standing near their roost entrance. They emerge from their roosts relatively slowly and generally, one at a time. The clicks of individual bats can be heard before they leave the roost which will then become silent, until the next bat prepares to emerge. They usually emerge later than other species.

It is important to differentiate the behaviour of Natterer's and Daubenton's bats over water. The Natterer's bat is usually seen flying higher, up to one metre above the surface, taking insects in the air. It is usually found over smaller bodies of water than the Daubenton's, where the sides are more thickly overhung with foliage. The Natterer's has a tighter turning circle and more abrupt changes in the repetition rate of the call, with longer periods of slower rhythm. Sequences of slower clicks are interspersed with a shower of very fast, short pulses. The Daubenton's flies parallel to the water surface, whereas the orientation of the Natterer's, to the surface, is continually changing. Daubenton's bats prefer to forage over calm water whilst Natterer's are frequently found hunting over riffled patches of water.

The "tik-ke-tik-ke" sound of the Daubenton's bat can be contrasted with the "tik-tik-tik" of the Natterer's.

Bechstein's bat (Myotis bechsteinii)

These bats are very rare. Their flight echolocation has been compared to the whiskered bat, but rather weaker. When leaving roosts, they are reported to use linear features such as forest roads and pathways to their foraging sites. Hans Baagoe noted a similarity, when hunting, to the Natterer's bat. They are also reported to feed by gleaning insects from tree trunks and foliage. Often feeding high up in the tree canopy, they are very agile. Bechstein's bats roost primarily in trees. **(Track 39)**

(Figure 19)

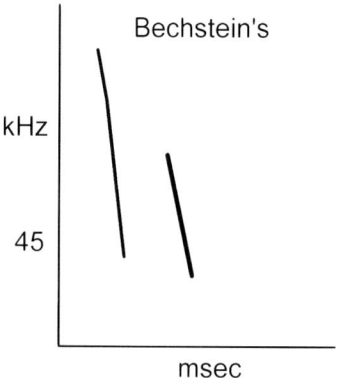

HORSESHOE BATS

These are the only two species in the UK that use constant frequency echolocation calls and their characteristic 'warbles', when heard on the bat detector, are unmistakable. Their distribution range is very limited, in the UK. When we visit areas which they inhabit, those of us who do not find them in our neighbourhood must remember to tune the detector higher than we would normally do, to listen at the frequencies above those used by the other UK species.

These bats derive their name from the shape of the flap of skin that surrounds their nostrils. Their echolocation calls are emitted through the nose and not through the mouth. The nose-leaf can be used to steer the stream of ultrasound rather like a moving torch beam.

Horseshoe bats are free-hanging bats and need to be able to fly directly through the roost entrance to their perch. Other British bats can land outside the roost entrance and crawl in. This means that horseshoe bats must have a roost entrance that is at least as wide as their wingspan.

Because of the directionality of their echolocation calls, they are very difficult to detect when foraging and so they are best heard next to their roost sites.

Greater horseshoe (Rhinolophus ferrumequinum)
This is the largest European horseshoe bat which has a slow, fluttering flight with short glides and it usually flies low. The greater horseshoe uses a constant frequency call at around 80 kHz, with a brief FM start and end. It sounds like a tuneful warbling on the heterodyne detector at 80 kHz which is the second harmonic of the bat's call. When listening close to the bat, the fundamental can be picked up at 40 kHz.
(Tracks 40, 41, 42 and 43)

(Figure 20)

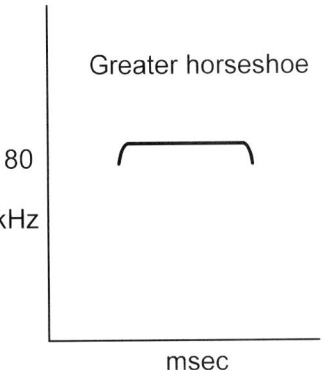

The 'Doppler shift' effect can be noticed as a bat flies past.

Lesser horseshoe (Rhinolophus hipposideros)
The lesser horseshoe also produces warbles of constant frequency but at 110 kHz. The pulses are shorter and the repetition rate is higher than that of the greater horseshoe.
The fundamental can be heard at 55 kHz when the bat flies near.

The higher frequency which this bat uses, makes its echolocation calls even more directional, making it much harder to detect, especially if the bat is flying in the opposite direction to the listener. **(Tracks 44 and 45)**

(Figure 21)

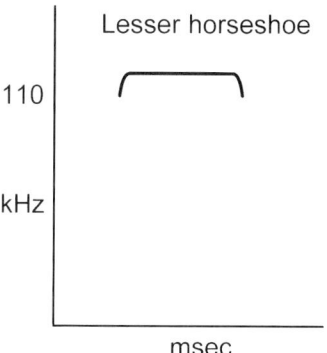

The FM at the start and end of each pulse can be heard as dry clicks at around 55 kHz.

As with the greater horseshoe, the 'Doppler shift' can be noticed as a bat flies past the observer.

FORAGING STYLES

A further aid to identification in the field is the recognition of how each group of bats catches its prey. A number of different strategies are used. Some species will use more than one method. Only the most frequently used are included here.

Hawking   Catching prey in the air, as used by: pipistrelle, serotine, noctule, Leisler's, whiskered/Brandt's, Natterer's, greater and lesser horseshoe.

Gaffing   Catching prey from the water surface: Daubenton's.

Gleaning   Taking insects from a surface: brown and grey long-eared, Bechstein's, Natterer's.

Stooping   Catching prey in flight, with a steep dive: noctule, Leisler's.

Pouncing   Taking insects from the ground: serotine, brown long-eared, greater and lesser horseshoe.

Perch feeding   Flying from a perch to catch prey in the air: greater and lesser horseshoe.

GETTING STARTED

The bat detector has a number of very important uses in the monitoring of bat activity. Apart from identification, these include the counting of bats from the roost, finding high activity centres and preferred feeding sites, tracking, measuring feeding rates and roost finding.

Now, let's get out into the field and start our search for bats!

Where do we begin and how do we stand the best chance of finding bats? We should concentrate on roost sites, commuting routes and foraging areas.

COMMUTING BATS will give the most species-specific echolocation calls. Listening to them, as they fly to and from feeding areas, gives the greatest chance of identification which can then be confirmed by netting at the roost, if necessary. Which may only be done by those possessing the appropriate licence.

Both summer and winter roost sites are significantly associated with water so a good place to start would be near to a stream, river or pond. We should look for areas with well developed linear features in the landscape, that interconnect with one another, with water and woodland. Here we should find the densest network of flight-paths.

Open areas with few interconnecting features have relatively low numbers of flight-paths.

Which features in the landscape are most used by bats?
It has long been recognised that bats are faithful to their summer and winter roost sites. Research, primarily from the Netherlands, has shown that bats are also faithful to their commuting routes. These are the features of the landscape which they use to navigate from roost to foraging area and back again. Such linear features are lanes, avenues, hedges, rivers, canals and woodland edges. A complex landscape allows a greater choice of routes and also seems to be a significant factor for enhancing bat activity.

The dependency of individual species of bat on these features is determined by the size of the bat and the range of its echolocation calls. Smaller species such as the pipistrelle, whose echolocation range is about 25-50 metres follow such features quite faithfully. Larger, louder bats such as the noctule will fly straight over the top of these small features using larger objects such as clumps of woodland or church spires as navigation 'beacons'.

Whilst a Daubenton's bat will primarily use a hedge or a woodland edge as a commuting route to its preferred foraging habitat over water, opportunistic feeders such as pipistrelles, brown long-eared and Natterer's will also feed along these commuting routes. Serotines use such commuting routes but do most of their hunting in more open areas.

Hedges and vegetation borders are rich in insects and provide shelter from wind and predators. Woodland edge extends in three dimensions to involve the canopy as well as the sides. Bats do not often use hedges less than one metre high and they usually fly on the leeward side.

Many bat species take part in **SWARMING** activity, where bats gather together and fly, sometimes in large numbers, just outside their roost on their return from foraging. Their flight, during this activity, is circular and repetitive, not unlike bees swarming. It can be very dramatic when seen at dawn, against a brightening morning sky. The activity can also be seen when bats return in the earlier part of the night, after their first foraging. The bats fly round and about the roost, in social display, sometimes calling to one another. Swarming may last up to three quarters of an hour. Some species that roost within open loft areas of buildings swarm within the building, but the typical swarming calls may still be heard outside on a bat detector. **(Tracks 47 and 48)**

Bats use linear features in a predictable way and this allows us to track them. Examination of the habitat, both on the ground and from maps, will predict the likely areas where roosts are to be found. Identifying a commuting route, in the evening as the bats leave the roost, enables us to track them again as they return to the roost, the following dawn

This skill has been the basic premise upon which most roost-finding surveys, using bat detectors, have been based.

It is beyond the scope of this book to describe the many different survey techniques in detail, but the system described in the next section, based on methods evolved during the Netherlands National Bat Survey, has proved its worth and is well within the capabilities of a local bat group.

A BAT DETECTOR SURVEY METHOD

Sites are visited three times each year:

- In the **Spring** - to find feeding areas, flight paths and make a **general assessment of species** in the area.

- In the **Summer** - again to find flight paths and feeding areas but primarily to identify **maternity colonies.**

- In the **Autumn** - flight paths, hunting areas, roost sites, **mating territories and mating roosts.**

A night in the field conveniently divides itself into three phases:

- **Sunset:** from 30 minutes before sunset until about 90 minutes after, is the best time to look for flight paths. Follow the bats to their feeding sites or briskly back-track to try to locate their roost.

- **Middle period:** searching for feeding areas, finding and identifying other species which emerge later.

- **Dawn:** from 90 minutes before dawn to about 30 minutes after, is the best time to track bats back to their roosts. Swarming activity can be seen easily, as the bats often continue swarming well after day-break.

Having predicted, from examination of the habitat, which species are expected to be present in an area, repeated visits should be made until hopefully, the likely bats have been located.

General area survey.
Walk or cycle the area, noting the species found, the time of observation and the direction of flight at sunset and sunrise. A cluster of bats flying at either of those times suggests that a roost is nearby.

Linear features.
Concentrate on linear features to find flight paths, particularly those that

lead to water, and advance in the direction from which the bats seem to be coming. On occasions, if fast enough, one can be led directly to a roost.

When all the bats have passed, record the numbers, species, time and direction. Pick up that trail again, towards dawn, as the bats return to their roost and follow them 'home'. If you have no luck, repeat the procedure for a few more nights, narrowing down the search.

Roost sites.

If there are no obvious leads, try to predict the likely area of roost site, from a knowledge of the species found and their preferred roosting habitat e.g. serotines in buildings and noctules in trees found in older parts of the woodland.

Stationed in the predicted roost area in the morning, you can expect the bats to concentrate more and more in one place, and look for swarming activity. Some species swarm within the roost and, occasionally, you may hear this on the detector, after the bats have entered.

Territorial males.

These may be heard calling from mating roosts, or in song flight, as they patrol their territory. Many of the mating calls are known to be species-specific. Listening with a time expansion detector can give extra information that can be used to identify the species. **(Track 26)**

Netting.

The identification of all Myotis species can be confirmed by netting at the roost entrance as the bats emerge.

NOTE: All bats and all bat roosts are protected by the Wildlife and Countryside Act (1981). No bat may be caught or roost disturbed without the possession of the appropriate licence.

The Bat Conservation Trust organises workshops to train people in these techniques. Their address will be found at the back of the book.

SUMMARY CHARTS

	Maximum loudness (Best listening frequency)	Sounds like
Greater horseshoe (CF)	80 kHz	warble
Lesser horseshoe (CF)	108 kHz	warble
Serotine	27 kHz	Tock
Pipistrelle (45)	45 kHz	Smack
Pipistrelle (55)	55 kHz	Smack
Nathusius's pipistrelle	39 kHz	Plop
Noctule	20-25 kHz	Chip-chop
Leisler's	25 kHz	Chip
Barbastelle	32 kHz	Tack
Brown long-eared	45-50 kHz	Tick
Daubenton's	45 kHz	Tik-ke
Whiskered/Brandt's	45 kHz	Tik
Natterer's	50 kHz	Tik
Bechstein's	50 kHz	Tik

Relative speed of Pulse-rate

SLOW	Noctule	**MEDIUM**	Whiskered	**FAST**	Long-eared
	Serotine		Brandt's		Daubenton's
	Leisler's		Pipistrelle		Natterer's

Most common summer Roost Sites

Pipistrelle	Buildings (often newer), bat boxes.
Noctule	Trees, rarely in buildings.
Leisler's	Trees, buildings more commonly than noctule.
Serotine	Buildings (often older).
Long-eared	Buildings (often older with slate roofs).
Daubenton's	Trees, bridges and other buildings, underground.
Whiskered	Buildings, trees.
Brandt's	Buildings, trees.
Natterer's	Timber framed barns, other older buildings, trees.
Greater horseshoe	Buildings, caves.
Lesser horseshoe	Buildings.

BAT DETECTOR SUPPLIERS

Stag Electronics: 15 Sir Georges Place, Steyning, West Sussex BN44 3LS.
Tel/Fax: 07000 228269 email: info@batbox.com www.batbox.com

Skye Instruments Ltd: Unit 32, Ddole Industrial Estate, Llandrindod Wells, Powys LD1 6DF. Tel: 01597 824811. Fax: 01597 824812.
email: skye.instruments@almac.co.uk

Ultrasound Advice: 23 Aberdeen Road, London N5 2UG.
Tel: 0171 359 1718 Fax: 0171 359 3650.

Pettersson Electronik AB: Tallbacksvagen 51, Uppsala, Sweden.
Internet: http://www.banhof.se/-pettersson/. Tel: 0046 1830 3880. Fax: 0046 1830 3840. email: pettersson@banhof.se

Daedalus: Unit G8, Greensmith Business Centre, Three Colts Lane, London E2 6BJ
Tel/Fax: 0171 247 7793.

David J Bale: 3 Suffolk Street, Cheltenham, Gloucestershire, GL50 2DH
Tel/Fax: 01242 570123

Magenta Electronics Ltd: 135 Hunter Street, Burton on Trent, Staffordshire DE14 2ST
Tel: 01283 565435 Fax: 01283 546932.

Maplin Electronics plc: PO Box 3, Rayleigh, Essex SS6 8LR.
Tel (sales): 01702 554161, (Enqs): 01702 552911. Fax 01702 445 935

A leaflet comparing the different bat detectors is available from The Bat Conservation Trust.

USEFUL ADDRESSES

The Bat Conservation Trust.
15 Cloisters House, 8 Battersea Park Road, London SW8 4BG.
Tel: 0171 627 2629. Fax: 0171 627 2628.

English Nature.
Northminster House, Peterborough, PE1 1UA.
Tel: 01733 340345.

Scottish Natural Heritage.
12 Hope Terrace, Edinburgh EH9 2AS.
Tel: 0131 554 9797.

Countryside Council for Wales.
Plas Penrhos, Fford Penrhos, Bangor, Gwynedd, LL57 2LQ
Tel: 01248 370444.

COMPACT DISC INDEX

Track 1. Two (artificially generated) constant-frequency tones (CF) at 1 kHz.

Track 2. Two slow frequency-modulated sweeps (slow FM).

Track 3. Three fast frequency-modulated sweeps (fast FM).

Track 4. Three frequency-modulated sweeps ending in near-constant-frequency (FM-CF).

Track 5. NOCTULE flying high above the tree tops, with the typical, loud, double-pulse call (chip-chop). The bat can be heard flying away from the observer and then flying nearer again, with some feeding buzzes. The 'wet' slaps at the tail of the near-CF call can be heard. (25 kHz)

Track 6. NOCTULE flying low, feeding along a hedgerow, with a much faster repetition rate than in track 5. The call has become a single pulse with virtually no near-CF component. (25 kHz)

Track 7. PIPISTRELLE. An isolated feeding buzz, recorded at 45 kHz (three times).

Track 8. PIPISTRELLE foraging, making many feeding buzzes. (45 kHz)

Track 9. SEROTINE. A single bat. Note the increase in the pulse rate as the bat prepares to catch the insect. The feeding buzz is emitted twice. (25 kHz)

Track 10. SEROTINE. The above recording slowed down fifteen times.

Track 11. BRANDT'S bat feeding. The pulses are fast and 'dry', without any hint of a 'wet' smack. This is in contrast to the pipistrelle bat that is heard near to the end of this track. (45 kHz)

Track 12. SEROTINE feeding at a street lamp. Note the 'wet' smack in the call, which is the characteristic sound when the near-CF tail is heard on a heterodyne detector. (25 kHz)

Track 13. DAUBENTON'S with PIPISTRELLE bats flying together, over water, the Daubenton's making 'dry' fast FM sweeps and the pipistrelle 'wet' near-CF pulses. these two species often forage together, the pipistrelle flying a few feet above the Daubenton's. Feeding buzzes can be heard.(45 kHz)

Track 14. PIPISTRELLE and DAUBENTON'S bats flying over water. (45 kHz)

Track 15. PIPISTRELLE. 'Typical' call, sounds *(stereo)* like clapping with cupped hands. Occasional social calls can be heard as short, dry buzzes at this frequency. (45 kHz)

Track 16. PIPISTRELLE flying high up in the tree canopy. Harder 'dryer' calls that can be confused with the sounds of Myotis bats but 'wet' near-CF tails can still be heard. (45 kHz)

Track 17. PIPISTRELLE foraging round the lower part of the tree foliage. (45 kHz)

Track 18. PIPISTRELLE flying in a very confined space. No 'wet' sounds can be heard at all. It is indistinguishable from a Myotis bat in these circumstances. (45 kHz)

Track 19. PIPISTRELLE 'gliding'. Sometimes pipistrelles will miss a few wing beats. (45 kHz)

Track 20. PIPISTRELLE returning to the roost; a rather 'fluffy' sound. (45 kHz)

Track 21. PIPISTRELLE. Tuning of the detector is offset by 5 kHz, producing a higher pitched sound. (40 kHz)

Track 22. PIPISTRELLE. Two individuals, flying together. Listening to the FM sweep of one and the near-CF tail of the other. (45 kHz)

Track 23. PIPISTRELLE. Feeding and social calls. (45 kHz)

Track 24. PIPISTRELLE social calls. (20 kHz)

Track 25. PIPISTRELLE social calls recorded with a heterodyne detector and slowed down ten times. Echolocation pulses sounding like a deep boom and the social calls as a slow, echoing trill.

Track 26. PIPISTRELLE social calls heard as a musical trill with 'time expansion' recording. The echolocation calls are heard as a high pitched 'peep'.

Track 27. BROWN LONG EARED bat, flying in a confined space (45 kHz).

Track 28. BROWN LONG EARED. A more typical recording from the field, the bat circling within a large barn. Brief snatches of very quiet calls. (50 kHz)

Track 29. SEROTINE. 'Typical' call; an irregular rhythm with a single tone. (25 kHz)

Track 30. SEROTINE recorded at 38 kHz, commuting close to a hedgerow. The detector is set above the frequency of the near-CF tail, so only the 'dry' sound of the FM sweep is audible. (38 kHz)

Track 31. NOCTULE and SEROTINE flying together, differentiated by the double-pulse of the noctule and the rhythm of the serotine. (25 kHz)

Track 32. LEISLER'S sounds similar to the noctule. The pulse-rate is usually faster and it uses single-pulse calls more frequently, at all heights, than the noctule. (25 kHz)

Track 33. BARBASTELLE. A complex call, with some 'wet' smacks and an alternating, fast and slow rhythm. (35 kHz)

Track 34. PIPISTRELLE and Myotis bats flying together. (45kHz)

Track 35. BRANDT'S bats emerging from their roost. Social noises from within the roost can be heard (45 kHz).

Track 36. DAUBENTON'S with fast, 'dry' FM calls, joined by a pipistrelle (45 kHz).

Track 37. NATTERER'S bats. Two individuals flying, one after the other (45 kHz).

Track 38. NATTERER'S. A large group of
(stereo) Natterer's bats foraging around a bridge over a riffled stream (45 kHz).

Track 39. BECHSTEIN'S bat recorded after release from a mist net. (45 kHz)

Track 40. GREATER HORSESHOE. Two bats in the roost, one flies off. The FM sweep can be heard as a dry click. (80 kHz).

Track 41. GREATER HORSESHOE. A group
(stereo) of bats flying near to the roost (80 kHz).

Track 42. GREATER HORSESHOE. Recorded as in track 41 but with the tuning offset. The FM sweep is more evident and the pitch of the calls is higher. (79 kHz)

Track 43. GREATER HORSESHOE. The fundamental frequency of the call (40 kHz). Most of the energy is put into the second harmonic at around 80 kHz.

Track 44. LESSER HORSESHOE. An individual. (110 kHz)

Track 45. LESSER HORSESHOE. A large group of bats flying within the roost. (110 kHz)

Track 46. PIPISTRELLE. '55' pipistrelles foraging at the water's edge; recorded at 45, 55 and 60 kHz.

Track 47. PIPISTRELLE bats swarming at dawn, outside their roost. Recorded at a distance of 30 metres from the roost (45 kHz).

Track 48. PIPISTRELLE bats swarming, at dawn, outside their roost. Recorded 3 metres from the roost. (45 kHz)

All tracks were recorded by David King using a Batbox III detector, except for the following:
Track 32 recorded by John Dobson using a Batbox III detector.
Tracks 12, 18, 24 and 48 by Brian Briggs using a Batbox III detector, track 6 with a Pettersson D980 and tracks 26, 28, 33, 38, 39 and 46 with a Pettersson D140.

FURTHER READING

Ahlen, I. (1990). Identification of Bats in Flight. Swedish Society for Conservation of Nature, Stockholm.

Altringham, J.D. (1996). Bat Biology and Conservation. Oxford University Press, Oxford.

Barataud, M. (1996). The World of Bats. Sittelle, Grenoble.

Catto, C. (1994). Bat Detector Manual. The Bat Conservation Trust, London.

Fenton, M. (1985). Communication in the Chiroptera. Indiana University Press, Bloomington.

Fenton, M., Racey, P.R. & Rayner, J.M.V. (eds). 1987. Recent Advances in the Study of Bats. Cambridge University Press, Cambridge.

Griffin, D. (1986). Listening in the Dark. Cornell University Press, Ithaca.

Hutson, A. (1993). Action Plan for Conservation of Bats in the United Kingdom. The Bat Conservation Trust, London.

Kapteyn, K (ed). 1993. Proceedings of the First European Bat Detector Workshop . Netherlands Bat Research Foundation, Amsterdam.

Limpens, H.J.G.H, & Kapteyn, K. (1991). Bats, their behaviour and linear landscape elements. Myotis 29: 39-47.

Richardson, P. (1985). Bats. Whittet Books, London.

Schober, W. & Grimmberger, E. (1989). A Guide to the bats of Britain and Europe (ed. R. Stebbings). Hamlyn, London.